Space Travel
Blast-Off Day

by Janet McDonnell
illustrated by Rondi Collette

Created by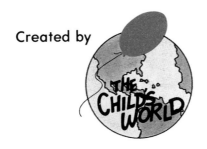

Distributed by CHILDRENS PRESS®
Chicago, Illinois

Grateful appreciation is expressed to Elizabeth Hammerman, Ed. D., Science Education Specialist, for her services as consultant.

CHILDRENS PRESS HARDCOVER EDITION
ISBN 0-516-08112-8

CHILDRENS PRESS PAPERBACK EDITION
ISBN 0-516-48112-6

Library of Congress Cataloging in Publication Data

McDonnell, Janet, 1962-
  Space travel : blast-off day / by Janet McDonnell ; illustrated by Rondi Collette.
    p. cm. — (Discovery world)
    Summary: Describes how astronauts aboard the space shuttle eat, drink, sleep, conduct experiments, walk in space, and feel when they arrive back on earth.
    ISBN 0-89565-556-X
    1. Space stations—Juvenile literature. [1. Space flight.
2. Astronauts.   3. Space shuttles.]   I. Collette, Rondi, ill.
II. Title.   III. Series.
TL797.M34   1990
629.45—dc20   89-23999
                                                  CIP
                                                   AC

1 2 3 4 5 6 7 8 9 10 11 12 R 99 98 97 96 95 94 93 92 91 90

Space Travel
Blast-Off Day

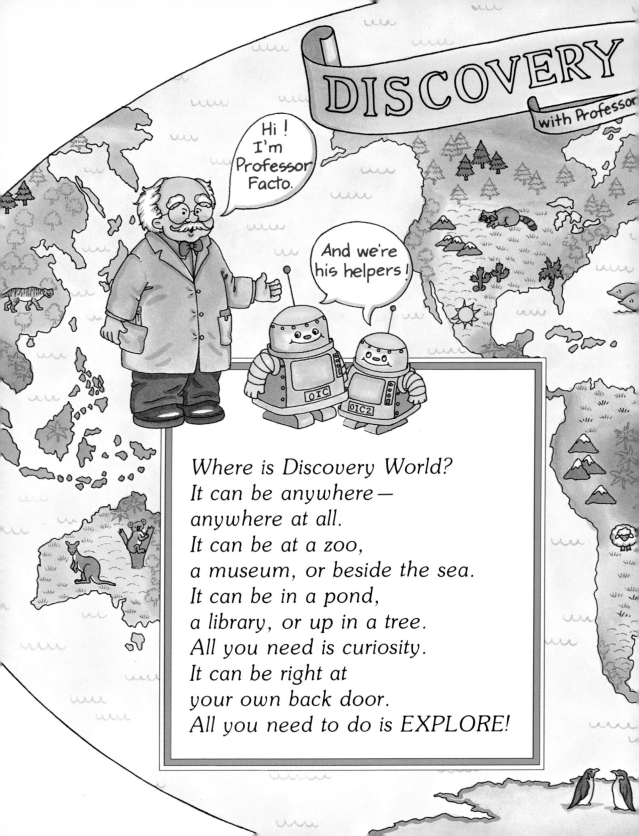

So come along and find out more about . . .

SPACE TRAVEL!

Today is the big day—blast-off day!
The astronauts are ready. They take an
elevator to the top of the space shuttle and
climb inside.

All strapped in, they wait for the count-
down. 5 . . . 4 . . . 3 . . . 2 . . . 1 . . .

Blast off! Two rockets take the shuttle
straight up, up, up, with great bursts of
fire.

In just a few minutes, the astronauts are in
space. From high up, they can see the
whole United States!

The space shuttle travels in a great big
circle around and around the earth.

While the shuttle circles the earth, the astronauts are very busy.

First they get used to floating in air! In
space, everything that isn't strapped down
floats—even the astronauts.

In the beginning they feel very strange, but soon they are having fun. They push off from the walls and do flips in the air.

Simple things are hard to do in space.
Even eating lunch is tricky. If an astronaut
lets go of his sandwich, it floats apart!

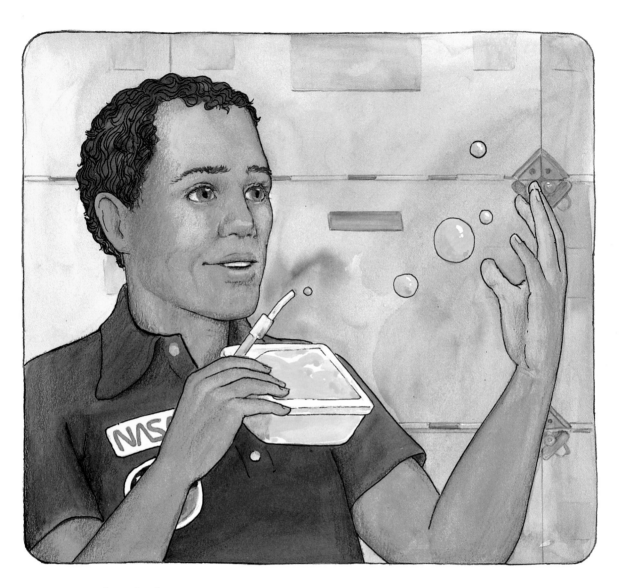

And the astronauts can't drink out of cups. The juice would just float right out. They drink from cartons with straws. If a drop does spill, it floats through the air in a little ball!

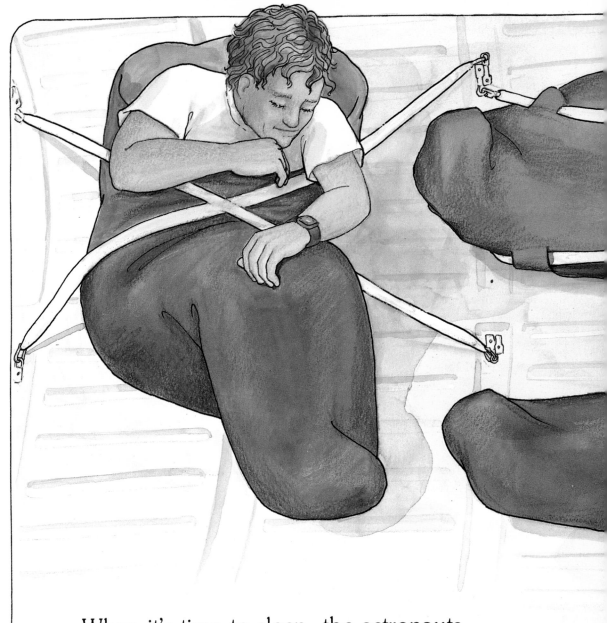

When it's time to sleep, the astronauts climb into sleeping bags. The bags are tied down so the astronauts won't float into each other as they sleep.

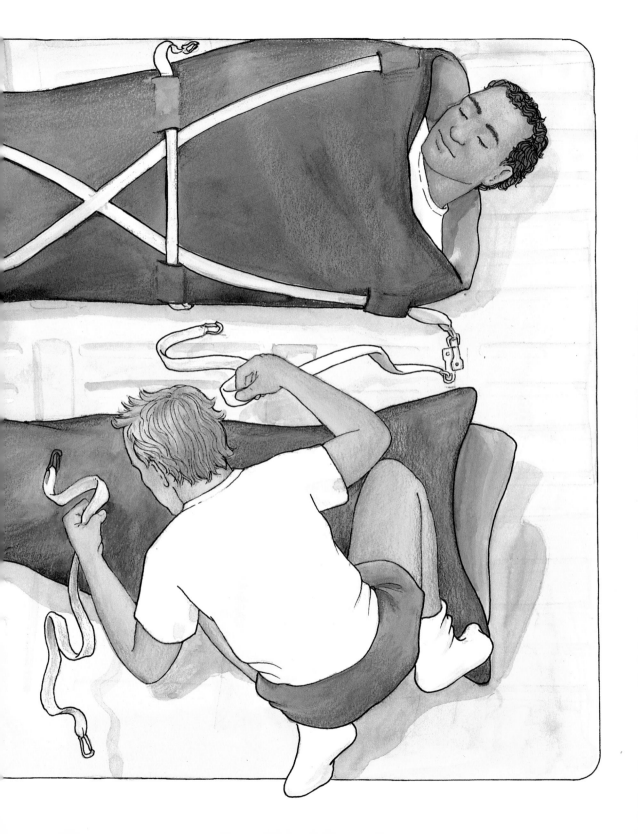

When they wake up, there's lots of work
to do. The astronauts have to check
experiments. One experiment shows how
a spider weaves its web in space. It seems
to be doing fine.

Another experiment is to find out if seeds
can grow normally in space. The pumpkin
seeds are doing fine too.

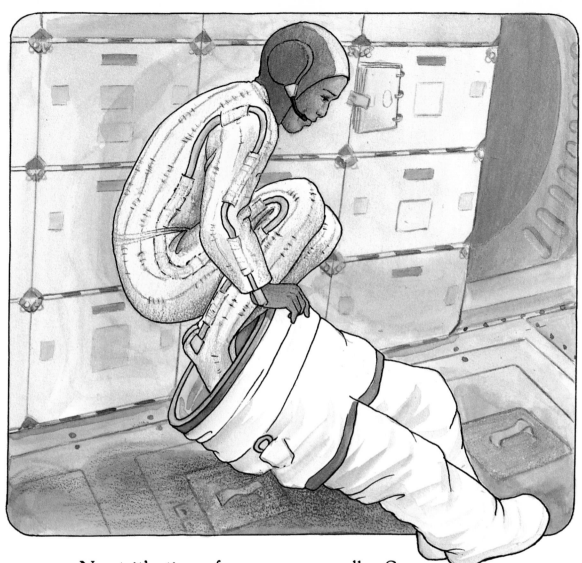

Next it's time for a spacewalk. Some
pieces of the shuttle were broken during
takeoff. Two astronauts must go out to
repair them. But first, they spend as much
as two hours putting on their spacesuits
and checking them to be sure they work.

Spacesuits keep the astronauts from getting
too hot or too cold. They also give the
astronauts oxygen to breathe. Without the
spacesuits, the astronauts would die in
minutes outside the space shuttle.

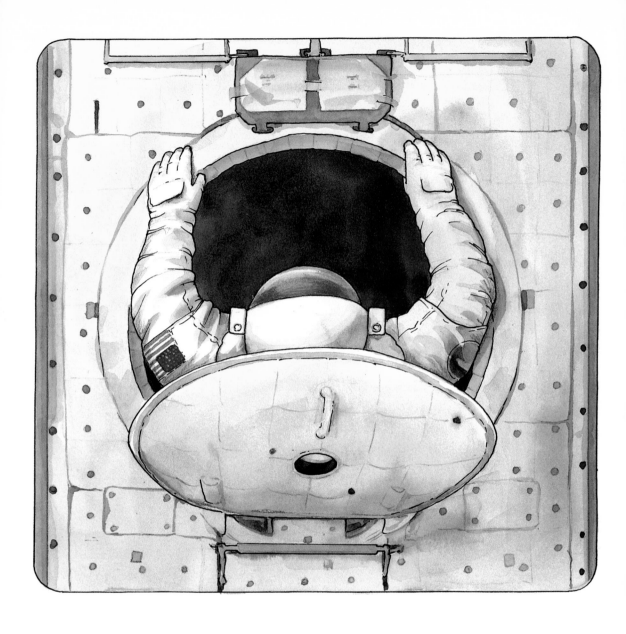

At last the astronauts are ready for their
walk in space. They leave the shuttle
through a special door.

Each astronaut is attached to the shuttle
with a line. They make their way to the
side of the shuttle. Then they go to work.

When all the astronauts' jobs are done, it's time to go home. The astronauts pack everything carefully away.

Then they strap themselves in and get
ready for the return trip. The shuttle slows
down and heads back to earth.

When the shuttle comes in for the landing, it is going 200 miles per hour! It lands on a special runway.

At first, the astronauts feel very strange to be back on earth. They feel as if they weigh a ton, and they can hardly lift their arms! But soon they are back to normal. It's great to be back home!

EXPLORE SOME MORE WITH PROFESSOR FACTO!

Sometimes when an astronaut takes a space-walk, he is attached to the shuttle with a line. But sometimes an astronaut has to take a walk in space far away from the shuttle. Then he can't be attached to it with a line.

Instead, the astronaut wears a jetpack. With the jetpack, the astronaut can zip right to where he wants to go. And when his work is done, he can zip right back to the space shuttle.

To see for yourself how a jetpack zips around, try making a balloon jet! Thread a 10 ft. (about 3 m) long piece of string through a straw. Have one person hold each end of the string. They should stand far enough apart that the string is pulled tight. (Or you can tape the ends to tables or chairs.) Blow up a balloon but do not tie the end. Carefully tape the balloon to the straw. Let go of the balloon and watch it go! How many puffs of air does it take to make the balloon jet go the whole length of the string? How many does it take to make it go half way?

Astronauts in the space shuttle travel about 200 miles into space. But some astronauts have gone by rocket much farther than that—all the way to the moon!

The first moon landing was in 1969. Three American astronauts made the journey. Astronaut Neil Armstrong was the first man to step on the moon.

On a later trip to the moon, astronauts brought a "moon car," called a lunar rover, with them. It was very heavy, so the astronauts left it on the moon when they returned to earth. It is still there!

On a clear night, go outside and find the moon. Draw what it looks like. Give two words that describe how it looks. Does the moon always look the same? How can you find out?

LET'S GO FOR A MOON RIDE!

OIC

INDEX

CHILDRENS PRESS

50495

9 780516 481128

ISBN 0-516-48112-6 NB21

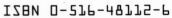

$4.

Sinbad
and the Roc

Cambridge
READING
Adventures

Ian Whybrow   **Nick Schon**

CAMBRIDGE
UNIVERSITY PRESS

University Printing House, Cambridge CB2 8BS, United Kingdom

Cambridge University Press is part of the University of Cambridge.

It furthers the University's mission by disseminating knowledge in the pursuit of education, learning and research at the highest international levels of excellence.

Information on this title: education.cambridge.org

Text © Copyright Ian Whybrow 2016; pedagogical content and reading guidance © Copyright UCL Institute of Education 2016
Illustrations © Cambridge University Press 2016

Series Editors (and authors of reading notes): Sue Bodman and Glen Franklin 2016

First published 2016

Printed in Dubai by Oriental Press

A catalogue record for this publication is available from the British Library

ISBN 978-1-316-50340-9 Paperback